# DINOSAURS

B. Mitchell

# Contents

# TRIASSIC DINOSAURS

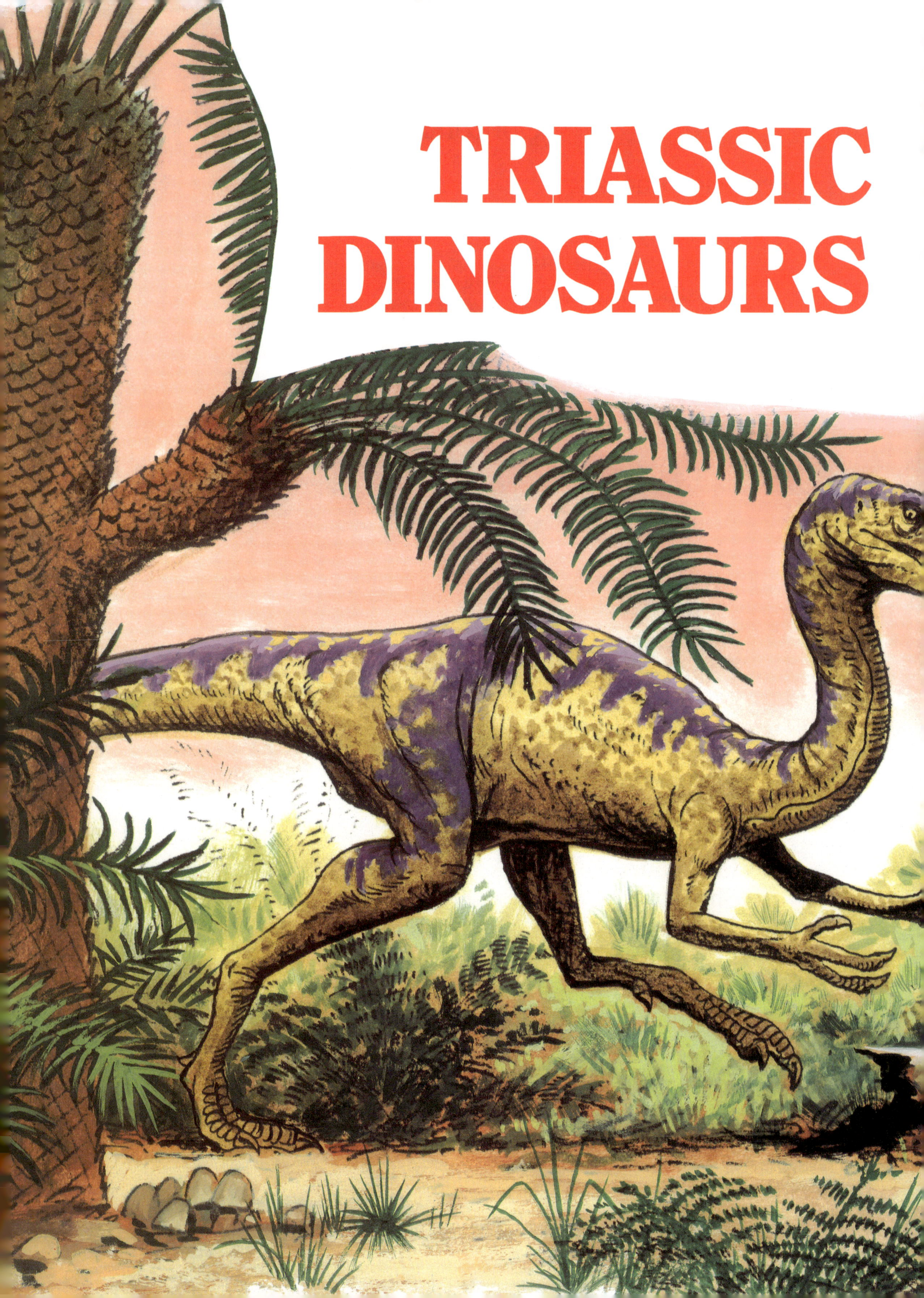

# Coelophysis

Coelophysis [see-lo-FISE-iss] was built for speed. This two-legged hunter had hollow bones, a long, flexible neck, a narrow skull, and a very long tail. Coelophysis ran fast, chasing lizards, dragonflies, small mammals and frogs. It caught these with its grasping, three-fingered hands. Coelophysis probably hunted in packs to prey on larger dinosaurs.

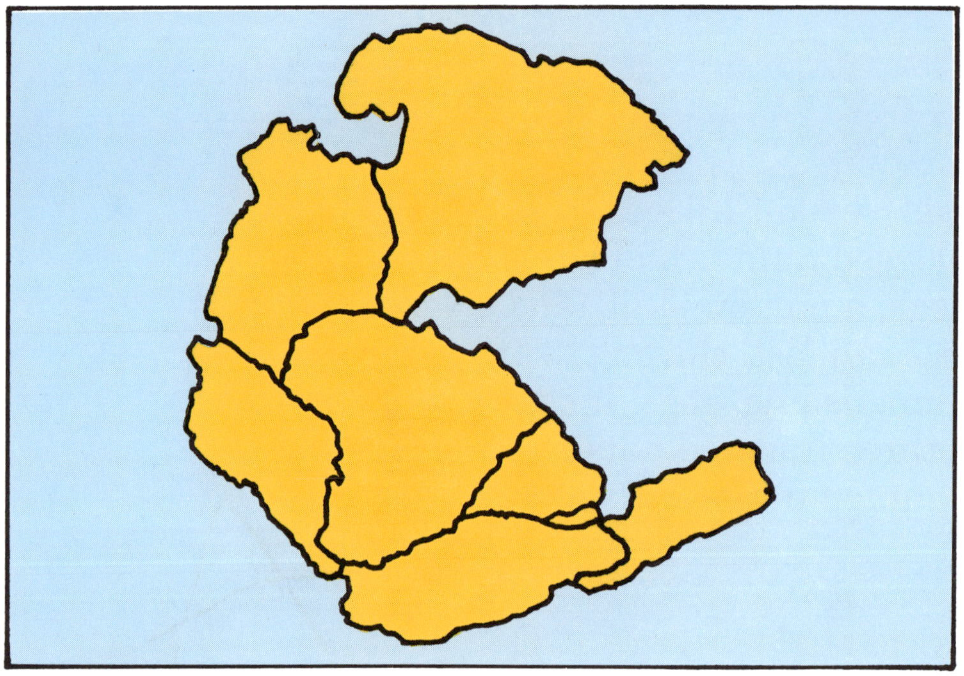

## Triassic Period

The Triassic [try-ASS-ik] Period lasted about 35 million years, until 190 million years ago. At the beginning all the land on earth was in one huge mass, crowded together on one side of the globe. The weather was mild, and there were huge rivers, dry deserts and great forests. Later, the land broke apart and formed two masses. The weather grew warmer and drier. Dinosaurs and mammals seem to have appeared about the same time.

## Procompsognathus

Procompsognathus [pro-komp-so-NAY-thus] was a small dinosaur with hollow bones and birdlike feet. It was 4 feet (1.2 m) long from its snout to the tip of its very long tail. Procompsognathus ran on its toes, leaning forward to grab insects and small creatures in its four-fingered hands. Much of what is known about the way this dinosaur walked and ran is from fossil footprint clues.

# Heterodontosaurus and Lesothosaurus

Heterodontosaurus [het-er-uh-DON-tuh-sawr-us] was a small, two-legged plant-eater with unusual teeth. It had three different kinds. It had sharp front teeth and a horn-covered beak for biting. Closely-spaced cheek teeth were for grinding. Here was a dinosaur that could chew!

Lesothosaurus [les-o-toe-SAWR-us] was a fast, two-legged plant-eater, only 35 inches long (89 cm), with light bones and leaf-shaped, jagged teeth.

# Ultrasaurus

Ultrasaurus [ul-tra-SAUR-us] probably managed to touch tree branches 60 feet (18 m) off the ground to nibble the topmost leaves. Herds of these enormous creatures thundered through forests, leaving a wide, cleared path. It may have been the largest dinosaur ever, possibly weighing 130 tons (118 t).

## Rhamphorhynchus

Rhamphorhynchus [ram-fo-RINK-us] had a long, slender head with sharp teeth. Its neck was long and could curve in an “s” shape. The rhamphorhynchus could snap its head forward like a spear, snagging fish on its long teeth. Its body and tail were 18 inches (45.7 cm) long and its wingspan was over 4 feet (1.2 m). A flat membrane on the end of its long stiff tail could have worked like a rudder to steer in the wind.

# Ichthyosaurs

Ichthyosaurs [IK-thee-uh-sawrs] swam in the early oceans. They looked like the porpoises of today with a smooth body and a long tail with a tail fin. Ichthyosaur had a neckless head, large eyes and a

long, thin snout lined with small, spiky teeth. Its front flippers were for steering through the water. Ichthyosaurs could not survive on land, even to lay eggs. Instead, the live young were born in the water, ready to swim.

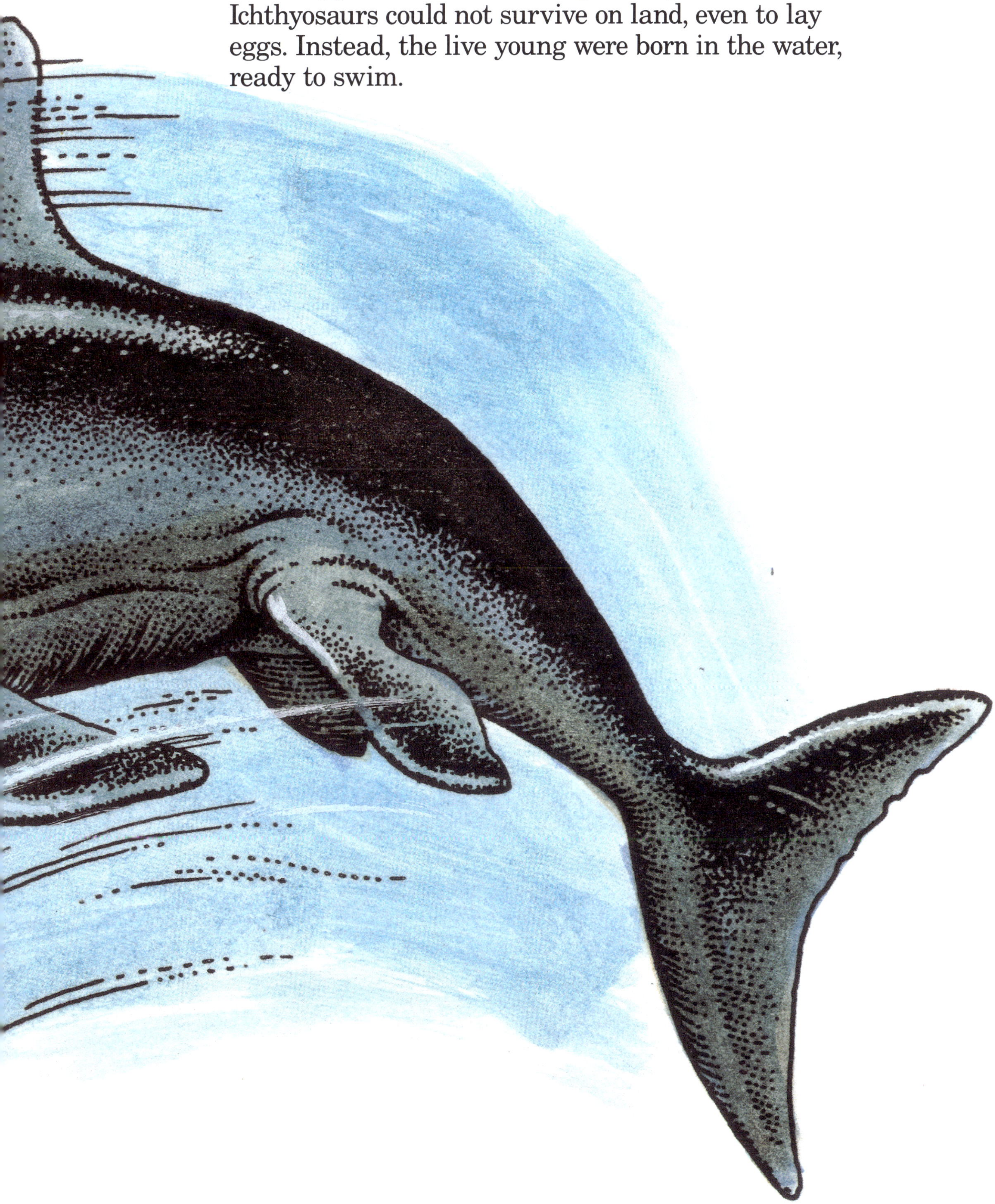

# JURASSIC DINOSAURS

## Stegosaurus and Allosaurus

Stegosaurus [STEG-uh-sawr-us] is best known for the row of leaf-shaped plates that ran down its neck, back, and part way down its tail. The biggest plates were often 2 feet tall (60.9 cm). It had a very small head for its body size.

Allosaurus [AL-uh-sawr-us] walked on two feet with powerful leg muscles. It could run as fast as a man, though it weighed 3 to 4 tons (2.7 to 3.6 t). Each foot had three sharp claws. Its arms were short, with three clawed fingers on each hand.

## Jurassic Period

During the Jurassic [jer-ASS-ik] period a great sea formed between the land that broke apart. Rain fell on the deserts. Land began to sink and water rushed in to form seas and lakes. Other land was pushed up to form mountains. The weather became warm and wet. Since there was plenty of food for plant-eating dinosaurs, they grew in numbers and became giants! As this period ended, 135 million years ago, the continents continued to pull apart.

## Archaeopteryx

Archaeopteryx [ar-kee-OP-ter-ix] was a remarkable winged creature that did not fly. Flapping like a chicken, it ran fast, snapping at flying insects. Archaeopteryx could grab an insect with its three-clawed hands and pop it into its tooth-lined beak. It is possible that the feathered wings held it in the air for a second or two as this first bird ran and leaped.

Mamenchisurus

Sichuan province in the People's Republic of China
is an especially rich area for "dragon bones."
Hundreds of skeletons have been unearthed, including
the remains of Mamenchisurus [mah-MEN-chee-sawr-us].
This 72 foot-long (21.9 m) plant-eater perhaps had
the longest neck of any dinosaur—33 feet (10 m) long!

# Allosaurus and Apatosaurus

Allosaurus [AL-uh-sawr-us] was a meat-eating dinosaur. It attacked the giant Apatosaurus and Brachiosaurus. It was 35 feet long (10.6 m) and tall enough to reach a second story window.

Apatosaurus [ah-PAT-uh-sawr-us] swallowed its food whole. It also swallowed stones that ground up the plant matter in its stomach. Apatosaurus had good eyesight and a good sense of smell. These helped it find and avoid its worst enemy, the ferocious Allosaurus.

## Camptosaurus

Camptosaurus [KAMP-tuh-saw-rus] walked on its hind legs, but often bent down on all fours to feed on plants. Solid arms, wide hands, and strong wrist bones helped support its weight while Camptosaurus cropped ferns. Grinding teeth were packed in the back of its jaw.

# CRETACEOUS DINOSAURS

## Triceratops and Tyrannosaurus

Triceratops [try-SAIR-uh-tops], or "three horned face," was the biggest horned dinosaur. It was 25 feet long (7.6 m), nearly 10 feet tall (3 m), and weighed about 6 tons (5.4 t). Triceratops had a short nose horn over its beak and two very long horns above each eye. These dinosaurs sometimes locked horns with each other to test who was the strongest.

Tyrannosaurus [tie-RAN-uh-sawr-us] walked or ran on two wide, three-clawed feet. It bent over with its tail straight out behind to balance its weight. Tyrannosaurus and Triceratops were the last dinosaurs to disappear from the earth.

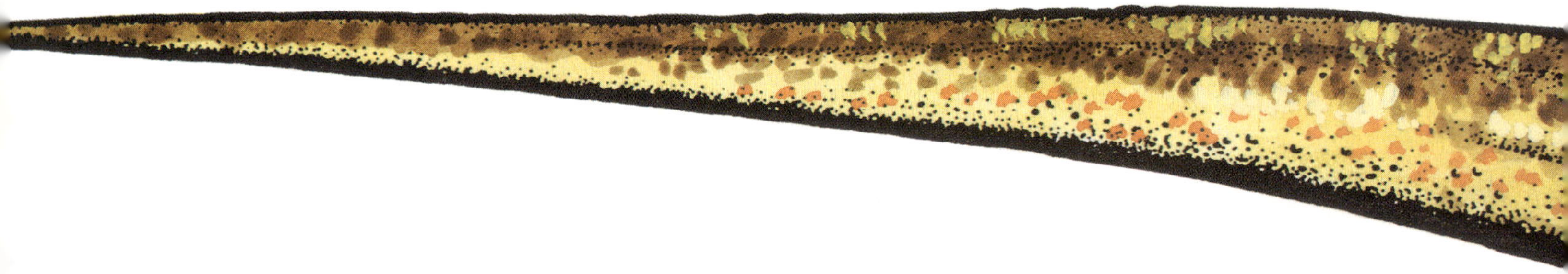

## Cretaceous Period

The Cretaceous [Kreh-TAY-shus] Period lasted 70 million years. Many continents of today formed during this time. The weather became seasonal—rainy and steamy or hot and dry. Flowers appeared for the first time! New kinds of plants meant new kinds of dinosaurs. The biggest animal ever to fly appeared during the Cretaceous and huge creatures lived in the inland seas. No one can be sure of what happened to cause all the dinosaurs to disappear, but this was the last era of the dinosaurs.

# Deinonychus

Deinonychus [dyne-ON-ik-us] was the most violent dinosaur of all time. It was very fast and had a new method of killing—with its feet! It stood about 5 feet tall (1.5 m), was 12 feet long (3.6 m) and ran on two toes of its strong, bird-like legs. The weapon on its middle toe was a 5-inch-long (12.7 cm), razor sharp claw used to slash its prey.

# Maiasaura

Maiasaura [mah-ee-ah-SAWR-uh], or "good mother lizard," dug their nests far apart, about the length of an adult maiasaura. Twenty softball-sized eggs were laid in a circle, then covered with leaves. When little peeping 13-inch-long (33 cm) dinosaurs pecked out of the eggs, the mother maiasaura uncovered the nest. The baby dinosaurs cheeped with their mouths open wide. They were fed with a mouthful of seeds and berries that their mother had already chewed.

## Spinosaurus

Spinosaurus [SPY-nuh-sawr-us] was 40 feet long (12 m) and weighed more than 7 tons (6.3 t). Its head was large and full of long, sharp, straight teeth. Huge spines stuck straight up along its

backbone, some as high as a tall man. These spines were covered with a sheet of skin to form a big fin in the shape of a sail.

## Parasaurolophus

Parasaurolophus [par-uh-sawr-OL-uh-fus] had the most amazing crest of all the dinosaurs. It was a long horn that started at the nose and extended back as much as 6 feet (1.8 m). A life-size model of a parasaurolophus crest was built. A low, loud honk, a sound somewhere between a goose and a fog horn, came out of the model.

## Tyrannosaurus

Tyrannosaurus [tie-RAN-uh-sawr-us] was the biggest meat-eating dinosaur. It had a simple hunting style. It opened its massive jaws and ran at its victim.

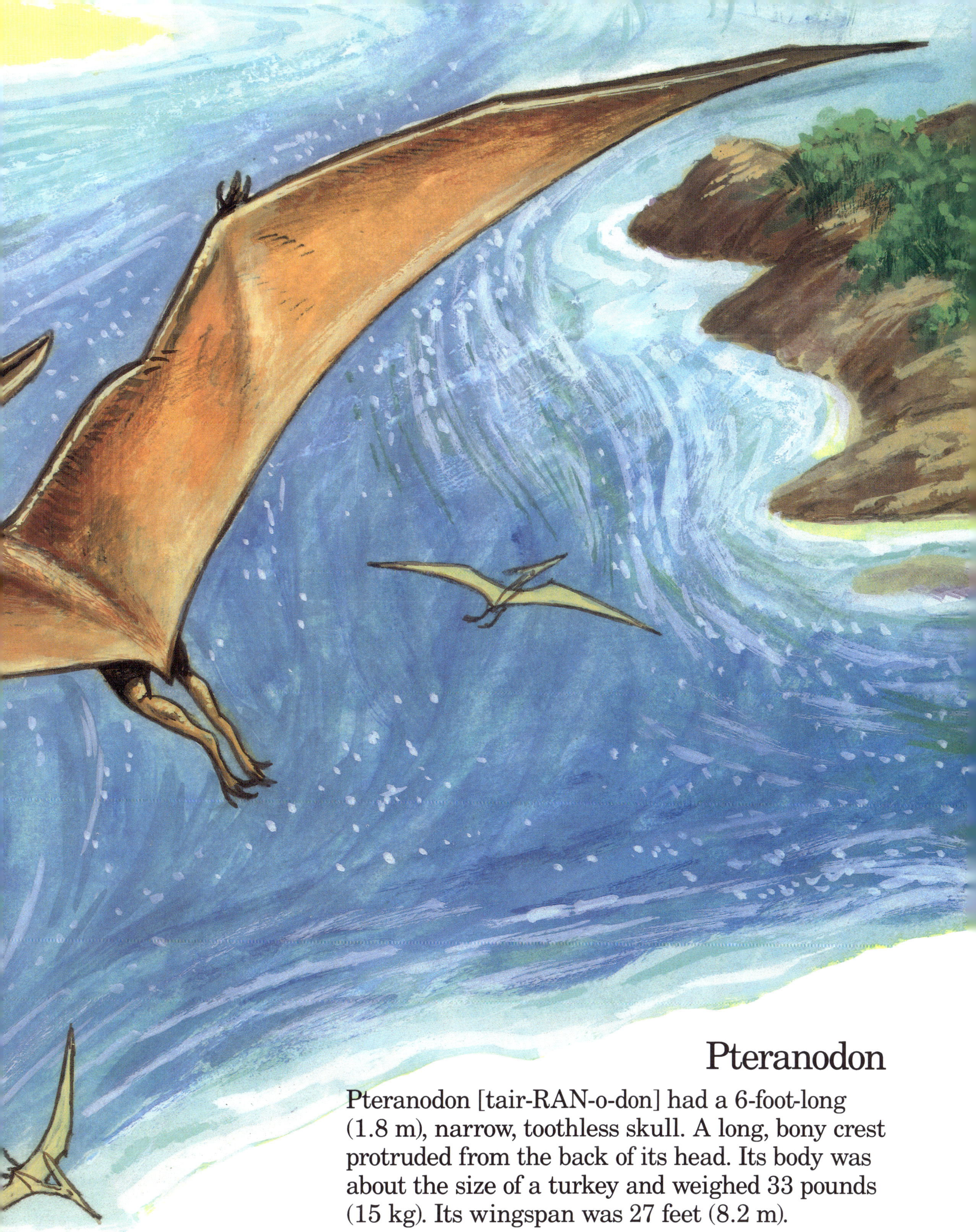

## Pteranodon

Pteranodon [tair-RAN-o-don] had a 6-foot-long (1.8 m), narrow, toothless skull. A long, bony crest protruded from the back of its head. Its body was about the size of a turkey and weighed 33 pounds (15 kg). Its wingspan was 27 feet (8.2 m).

## Pachycephalosaurus

Pachycephalosaurus [pak-ee-SEF-uh-lo-sawr-us] belongs to a group of dinosaurs called "boneheads." These dinosaurs had built-in crash helmets! The bone on top of their 2-foot-long (60.9 cm) skull

formed a dome nearly 10 inches (25 cm) thick. It was fringed with bony knobs and short spikes. Boneheads needed thick skulls to protect their brains from head butting.

# Iguanodon

Iguanodon [ig-WAN-o-don] was 25 feet long (7.6 m) and stood 12 to 15 feet tall (3.6 to 4.5 m). Iguanodon's skull was long and flat. It walked on two three-toed legs. It had a sharp thumb. Iguanodon was a plant-eater with a sharp beak and cheek teeth for grinding.

# Dr. Gideon and Mrs. Mary Ann Mantell

The first dinosaur bones to be studied by scientists belonged to Iguanodon. In 1822, Mary Ann Mantell picked up some unusual fossil teeth from a rock pile by the side of a road in Sussex, England. She showed the teeth to Dr. Mantell. He traced the rock pile to a nearby quarry and asked the quarrymen to bring him any fossils they uncovered. More bones and a horn were found.

Mary Ann Mantell

Dr. Gideon Mantell

# Oviraptor

Oviraptor [o-vee-RAP-tor] looked like a bird. Its bones were hollow, it had a toothless beak and it ran on two long, slender legs. Instead of wings, Oviraptor had long, thin arms and very large hands. Each hand had three fingers with sharp, curved claws. It appears that Oviraptor slipped into nests to steal eggs when parent dinosaurs weren't looking.

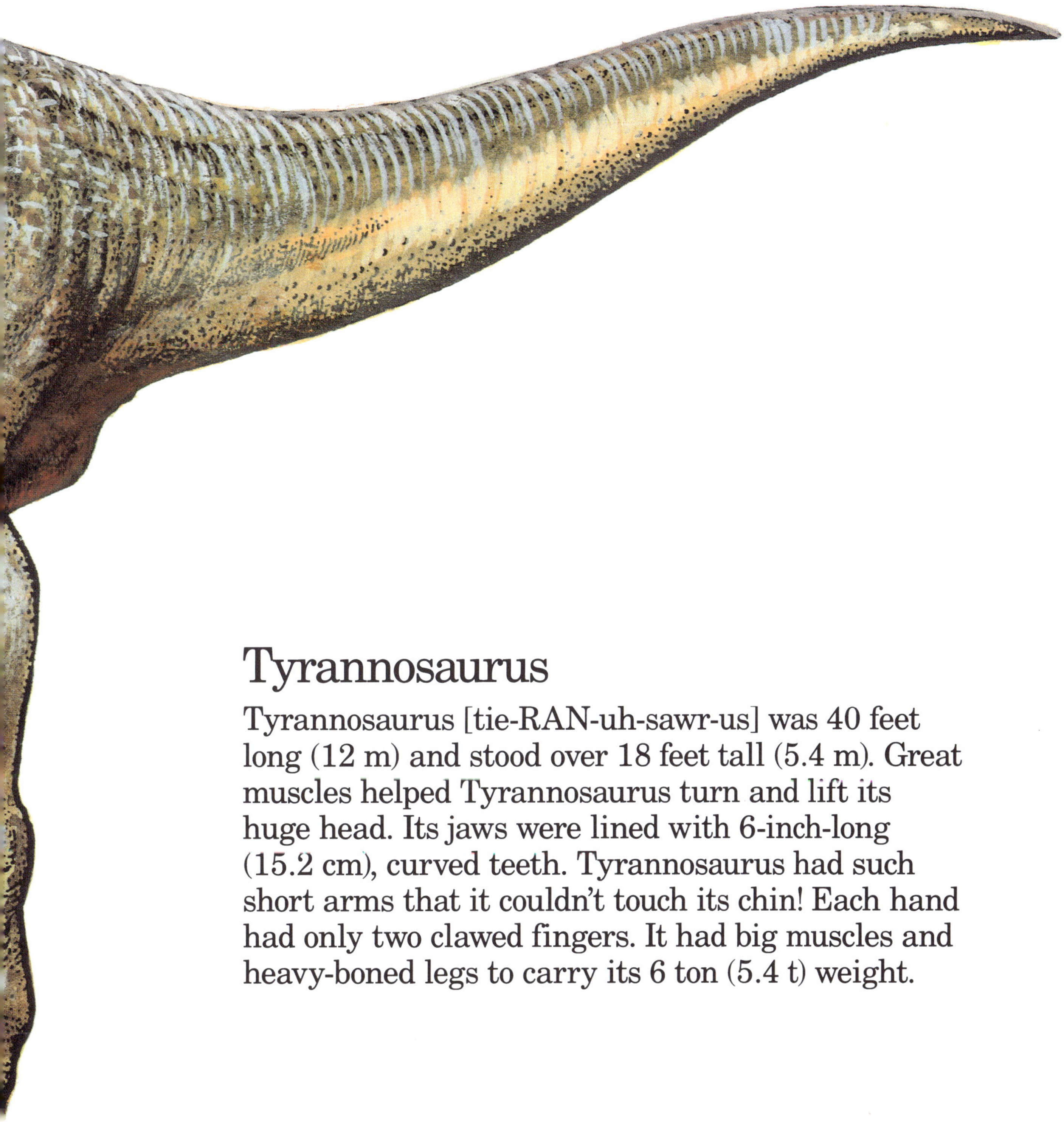

## Tyrannosaurus

Tyrannosaurus [tie-RAN-uh-sawr-us] was 40 feet long (12 m) and stood over 18 feet tall (5.4 m). Great muscles helped Tyrannosaurus turn and lift its huge head. Its jaws were lined with 6-inch-long (15.2 cm), curved teeth. Tyrannosaurus had such short arms that it couldn't touch its chin! Each hand had only two clawed fingers. It had big muscles and heavy-boned legs to carry its 6 ton (5.4 t) weight.

# GLOSSARY

Cretaceous Period—third period in the Mesozoic Era, 135 to 65 million years ago

dinosaurs—a group of land reptiles that appeared about 215 million years ago and became extinct about 65 million years ago

fossils—remains of prehistoric plants and animals that have turned to stone

Jurassic Period—second period in the Mesozoic Era from 200 million years ago to 135 million years ago

Mesozoic Era—period of time from 225 to 65 million years ago

Triassic Period—first period in the Mesozoic Era from 225 to 200 million years ago